SERVICE CHARGES IN
COMMERCIAL PROPERTY

SERVICE CHARGES
IN COMMERCIAL PROPERTY

Michael Young ARICS

with a Foreword by
T.B. Stapleton M.Phil, BSc, FRICS

Routledge
Taylor & Francis Group
New York London

First published 1992 by Estates Gazette

Published 2014 by Routledge
2 Park Square, Milton Park, Abingdon, Oxon OX14 4RN
711 Third Avenue, New York, NY, 10017, USA

Routledge is an imprint of the Taylor & Francis Group, an informa business

ISBN 13: 9780728201736 (pbk)

CONTENTS

TABLE OF CASES

FOREWORD

The practical nature of service charges and their interrelationship with accounting and landlord and tenant law renders the topic a particularly difficult one to be taught on undergraduate courses. Michael Young has however produced a readable and logically ordered monograph on the subject, which should enable those in other disciplines to understand the various aspects of service charges and those for whom it is their own discipline to have an accessible source of reference.

The discussion of accounting procedures in Chapter 4 is perhaps the particular contribution he makes to the development of the subject, and the text is as relevant to the in-house property manager as to the property management department in private practice.

The land steward of the Middle Ages still exists in the form of the modern property manager. He is likewise disciplined by his accountability to his client while recognising that such accountability means not just maintaining an effective financial regime but also having a real understanding of the client's needs and the skills to monitor their achievement.

T.B. Stapleton

INTRODUCTION

The aim of this book is to offer a practical guide to the often maligned and perhaps misunderstood subject of service charges in commercial property. Separate considerations apply to residential charges, which are subject to statutory regulation in certain circumstances (see ss. 18-30 of the Landlord and Tenant Act 1985 as amended by the Landlord and Tenant Act 1987).

Service charge provisions are well established in modern commercial leases, since the British property market demands that the landlord should receive 'clear' rent from multi-occupied buildings. The running costs of such buildings, or of estates, are separately charged to the tenant, and, therefore, the level of those charges can be an emotive issue, especially where potential exists for both landlords and tenants to misunderstand each other's objectives.

The principle of paying a service charge has gradually become more acceptable to occupiers, who have recognised the benefits of a well-maintained and well-managed building. A higher expectation of the standards of maintenance and the working environment in commercial property has developed, although mainly in the context of a rising market and an increase in wealth generally.

Most new multi-occupied buildings have been sufficiently well designed, and the leases suitably drafted, to ensure that the parties to the lease should be clear in their understanding of their obligations to each other. Problems arise, however, with older buildings, not designed to provide the level of services now expected, and/or with unsophisticated lease provisions accompanying an older tenancy.

Inherent in the misunderstanding of service charges is the fact that no two buildings, and thus no two leases, are exactly the same. It is hoped that this publication will help the practitioner to decide what are the most appropriate services and service charge provisions for any individual building.

In 1985, T.B. Stapleton outlined his thoughts on the direction in which service charge matters might develop ('Briefings for Surveyors, no. 5 – Service Charges', Henry Stewart Publications and Portsmouth Polytechnic). It is interesting to note what progress has been made, on the points he mentioned.

He foresaw:

1. a model service charge clause, built upon the experience of a model rent review clause (the RICS Guidance Notes are moving in that direction);
2. greater reference to actual service charge cost at rent review and lease renewal;
3. more analysis of service charge cost and hence pressure on property managers to explain variations from the norm;
4. greater emphasis on the independent arbitral function;
5. clearer emphasis on service charge cost in the negotiation of new lettings;
6. greater application in the monitoring of services by microtechnology;
7. alternatives to using a sinking fund being devised;
8. a steady reversal in the previously predominant position of landlord.

While greater attention has been paid to most of these matters, there have been no definitive changes. However, the late 1980s generally saw a rising market as a result of greater demand from occupiers following increased profits. Economic constraints, such as those of the early 1990s, have caused all occupiers to examine their costs closely, and, as service charges, therefore, come under closer scrutiny by tenants, especially the more sophisticated ones, the pace of change could accelerate. Well-informed and well-prepared landlords and property managers, owners and agents should, however, have little to fear.

1 SERVICE CHARGES – DEFINITION AND SCOPE

What is a Service Charge?

Service charges can be defined as the costs incurred by the landlord of any multi-tenanted property in maintaining and managing the range of activities and obligations related to the physical condition, environment, and use of both the demise and the common areas and facilities, where such costs are rechargeable to the tenant by virtue of the lease terms.

The phrase 'by virtue of the lease terms' is crucial. The tenant is obliged to pay only on the basis of the terms set out in the lease, subject to any modification by statute or by legal precedent – and thus it is essential to ensure that the lease is suitably drafted and clear.

The Contractual Basis

Modern forms of service charge clauses usually set out the obligations of both landlord and tenant. The tenant is specifically required to repay the landlord the cost incurred by the landlord in providing a clearly defined set of services, usually set out in a schedule to the lease and having regard to the definition of the demise, and the building of which the demise forms part. The following is an example of such a clause:

AND ALSO PAYING by way of additional yearly rent:
(i) at all times throughout the term a contribution ('the service charge') representing a proportion of the costs and expenses incurred by the Lessor in connection with the works services and other matters done or provided

by the Lessor which shall include the items set out in the
First Part of the Third Schedule hereto such contribution
to be calculated and payable in the manner set out in the
Second Part of the Third Schedule hereto PROVIDED
ALWAYS that the Lessor shall take all proper steps to
ensure that such costs expenses and outgoings are
properly incurred and are not unreasonable.

The landlord does not necessarily covenant to provide the
services, but most prudent tenants do usually require such a
specific covenant so that doubts as to obligations are minimised.
If there is no specific covenant, it would appear that the tenant
could not force the landlord to provide the service. However, it
would be counter-productive for most landlords not to do so
since it is their investment which needs presumably to be kept in
good order at all times so as to maintain, if not enhance, the
value.

It is also important for both parties to ensure that the extent
of the building to which the services pertain is clearly defined
within the lease, as well as the extent of the tenant's demise. In
this way, there should be no misunderstanding as to the respec-
tive obligations of landlord and tenant, nor as to the parts of the
building or the estate for which the common services are
provided, as distinct from the tenant's own demise where he may
be directly responsible for costs arising.

It is usually prudent to ensure that the lease stipulates the
times during which the services will be provided, for example,
in a multi-tenanted office building, from 7 a.m. to 7 p.m. If any
tenants require services, such as heating and security, outside
these hours, then they should be required to pay specifically for
those additional costs, as it is they who receive the direct benefit
and not the other tenants in the building who are not using the
facilities out of hours. This provision will be particularly relevant in office buildings where there is a wider range of services
common to the whole, but it will also apply in shopping centres,
where traders may have varying opening times (including Sun-
days) and servicing times and thus will use the services to

varying degrees. Such a provision may not be necessary in the context of industrial estates or business parks.

Some occupiers may, however, demand twenty-four-hour access to buildings. The foregoing provisions are designed not to fetter such access but merely to make it clear to those occupiers that they must pay for the services from which only they benefit. Usefully, in some buildings, building management computers are able to monitor more accurately the use of services.

The Basis of Payment

Modern leases usually require the tenant to make interim payments on account (known as 'advance payments'), at each usual quarterday, with provision for an annual reconciliation of the account where the total costs for the year will be certified (see below) and the tenant's proportion of those total costs will be set out in a statement. The advance payments will be deducted from the total due for the year, and a balancing sum either demanded from the tenant or repaid to it. In practice, the balancing adjustments are usually made at the next convenient quarterday, although leases quite often require these to be paid on demand – see sample provisions in Appendix 4.

Advance payments can also be demanded half yearly or, more rarely, annually in advance. The advance payment system is most common where the number and level of services provided are substantial, and this enables the landlord to fund those services without having to meet the costs from his own resources or rent.

Where services are relatively minimal, it is quite common for tenants to be charged quarterly or half yearly in arrears, or even, sometimes, on demand. This is obviously to the landlord's financial disadvantage, as he will have to fund those services until the costs are recouped from the tenants. Older forms of lease may specify that the landlord is only obliged to provide the services if the tenant is making the appropriate contribution to his costs. This is not particularly helpful where the costs are rechargeable in arrears and is an example of the kind of poor drafting that can lead to dispute.

In general, legal textbook writers are of the opinion that service charges should be payable as additional rent rather than there being a mere covenant to pay. This is because of the merits of distress as a means of enforcing recovery, rather than the use of forfeiture proceedings and the need to have regard to the requirements of s. 146, Law of Property Act 1925. The topic of distress is covered further below (see pp. 37-8). There is also the advantage that service charge 'rent' can be protected by insurance cover against loss of rent.

Older Lease Clauses

Many older forms of lease contain only common contribution clauses, such as:

> to pay a fair proportion of the costs incurred by the landlord in repairing rebuilding cleansing all party walls fences sewage drains conduits and other things the use of which is common to the demise and other premises.

Such a clause is not, strictly speaking, a service charge provision, and it is very doubtful whether many items normally included in modern service charge provisions would be recoverable by it, particularly since, as a general rule, being vague, it would be interpreted narrowly and in the tenant's favour. It would anyway be quite unsuitable for a building in multiple occupation and was probably designed to be used in the lease of a building within a terrace of similar properties all of which happened to be owned by the same landlord, for example, a parade of shops, which would otherwise be held individually on full repairing and insuring terms, but have certain basic facilities such as party walls and drains in common.

Apportionment of Costs

Common methods of apportionment and their applications are set out below. Practitioners should always check with leases to ensure that the apportionment of costs is carried out strictly in accordance with the method stipulated in the lease. Failure to follow lease procedure could result in the landlord being denied the ability to make due recovery.

Fixed Percentages

These have the merit of certainty and equity where the total adds up to a hundred per cent only, and where there is provision to alter the percentages should circumstances change, for instance, if additions are made to the building.

Fixed percentages are probably most appropriate in modern office buildings, where there is little variation in the extent of services and cubic capacity between the individual demised areas. Varying percentages can, however, be applied to those services which differ, e.g. heating. The tenants of an older building, for example, might argue that the higher ceilings and larger rooms on the lower floors of the building necessitate a greater consumption of heat than the smaller rooms on the upper floors. Therefore, a volume-based percentage could be applied to the tenant's heating contribution, with a floor-area-based percentage applied to all other services. This creates a two-tier service charge which can be easily apportioned and this is not uncommon. Modern computer-based service charge systems can deal easily with the concept of different percentages being applicable for different services.

From the tenant's point of view, potential disadvantages of specific percentages could arise in some circumstances. For example, a landlord who is occupying part of the building might see an advantage in apportioning the total charge among the occupants of the remainder of the accommodation, thus over-burdening them with costs. Similarly, specific percentages without a variation provision are inappropriate in buildings or estates where there is phased development, or potential redevelopment, or where occupation varies. To be equitable, therefore, recalculation procedures need to be set down in the lease.

It is good practice to advise prospective tenants as to the basis on which the percentages have been calculated, i.e. by reference to floor area or otherwise. Tenants will be anxious to ensure that their percentage does relate to their proportion of the whole building and that they will not be liable for any costs attributable to vacant areas. This is easy to demonstrate and draft into lease provisions. Clearly, untenanted areas are part of the landlord's risk and, as such, a fair system will ensure that the landlord bears all the costs of void areas – including the service charge.

Floor Area Apportionment

The lease usually stipulates that the apportionment of the floor area should be calculated according to the ratio of the individual floor areas (of the demise) to the total lettable floor area of the building, i.e.

$$\frac{\text{Individual area}}{\text{Total lettable area}} = \text{Proportion to be charged}$$

An example of this is as follows:

$$\text{Service charge cost } £100,000 \times \frac{\text{Total floor area 4,000 ft}^2}{\text{Total area 98,000 ft}^2}$$

$$= £4,081.63 \text{ tenant's contribution}$$

This method is both simple and popular, and entirely equitable when accompanied by a recalculation basis should circumstances change. Leases need also to specify that the measurement should be done in accordance with an agreed measuring code, e.g. gross internal floor area as defined by the RICS/ISVA

Code of Measuring Practice. Again, where services vary between different parts of the building or estate, a two-tier service charge can be introduced.

This basis of apportionment can be applied to all types of property but its disadvantage is that it will not reflect any difference in the quality of accommodation. There is a presumption in most service charge clauses that the accommodation is similar throughout any building; this is not necessarily the case in older buildings. However, this argument could be countered by the presumption, if not the fact, that the rent payable will reflect the quality of accommodation.

Rateable Value Apportionment

Apportionment on a rateable value basis usually overcomes the problem of disparity of accommodation, as this should be reflected in the rateable values. Apportionment by this method is generally on the basis of the proportion which the rateable value of the demise bears to the aggregate of all the rateable values in the building or the estate. Since rateable values change with each rating revaluation, the apportionment will need to change similarly. Indeed, the courts have held that the ratio of rateable values to be adopted must be that from time to time in force and not forever frozen at the ratio current when the lease was originally completed. This principle was expounded in *Moorcroft Estates Ltd. v Doxford* (1979) 254 EG 871. Furthermore, the new proportions must take effect as from the date at which the change in rateable values takes effect, which, of course, does not necessarily coincide with the service charge accounting year.

In a complex, new, multi-tenanted building such as a shopping centre, several years can pass before correct rateable values are agreed, and these can often vary substantially from the original assessments. These can then be backdated to the original date of occupation and, as a consequence, when rating assessments are finally settled, in equity, service charges need to be recalculated on the basis of the adjusted assessment. By this time, tenants may have left or assigned leases. This form of apportionment is not, therefore, always the most appropriate.

Volume Apportionments

These have been briefly mentioned above (p.5): they are not particularly popular or widespread. They are, however, useful where the cubic capacity of the demised areas varies considerably and are probably best expressed in percentage terms in the actual lease; otherwise, there is no reason why the basis on which the percentages are calculated cannot be by volume.

Site Area Apportionment

This method of apportionment is particularly relevant to large business parks. The principles of apportionment are the same as for floor areas within, say, an office building, but translated to the context of a business park with its large-scale landscaping and open spaces. Such developments are usually phased, and site areas can therefore vary in terms of the proportion of the developed area and of the whole site that they represent.

The site area apportionment method is used to calculate an individual freeholder or leaseholder's contribution to the cost of running services, such as landscaping, road maintenance, common security, signage, management, and other maintenance, which are common to the whole estate. In order to ensure continuity of maintenance, the landlord, as developer, is likely to retain, either directly or via a management company, the right to maintain all landscaping around the common areas.

Within each site demise, there could be a building in single or multiple occupation. If it is in multiple occupation, the tenants within that demise will bear a floor-area proportion of the services supplied to the building in much the same way as a normal office building, together with the immediate landscaping and car-parking areas accompanying the building. To this floor area apportionment will be added the site's contribution (on a site area basis) to the cost of the estate's common areas. This gives a two-tiered service charge.

The Services to be Provided

Services, usually set out in a schedule to a lease, will vary according to the building or estate, its common needs, and the type of use. Each building has, therefore, to be assessed individually. A typical multi-let office building, for instance, could include the following services:

Insurance
Repair and renewals to common areas (external + internal)
Heating, ventilation, air conditio.ing, all mechanical and
    electrical equipment, and fuel
Electricity to common areas
Cleaning and cleaning materials
Window cleaning
Floral and plant displays
External landscaping
Lift repairs and renewals
Fire-alarm systems and fire-fighting equipment
Water charges
Rent, rates on staff accommodation (or local tax)
Building staff – manager, porter, receptionist
Managing agent's fees or landlord's costs
Consultants' fees – engineers, surveyors, accountants,
    auditors
Cost of compliance with statutory requirements
Creation of a reserve fund
Cost of enforcing covenants against other tenants
Security, including car park and building access system

A shopping centre may have all these services and more, including promotion and marketing, and the cost of running a traders' association. An industrial estate will probably have considerably fewer common services. Typical ones include:

Road repairs
Landscape maintenance

External common areas – cleaning and litter collection
Drainage maintenance
Drainage pumps
Common lighting
Security
Caretaker
Common signage
Management fees
Consultants' fees

Business park services will encompass both office and industrial
items, weighted according to the control granted to the landlord.

2   IMPLEMENTATION

The owner of a commercial property investment usually has the ultimate right to decide which individual organisation will implement the management of the building and the service charges. In rare circumstances, the leases will actually specify what type of organisation should manage, e.g. 'a local firm of respectable managing agents and surveyors'. Thus the owner usually has, broadly, three options, each of which may have advantages.

Direct Management by Landlord

Some landlords prefer to deal with the management of their own properties direct or in-house. This applies to many of the larger institutions and property companies, who feel that they require more direct control over what is happening at their properties, particularly in terms of cash flow and income. In recent years, however, the desire for increased competitiveness and productivity has caused landlords to examine the cost-effectiveness of continuing to manage the property directly themselves, and the trend has been for landlords to continue to collect rent but to put the day-to-day management of buildings and service charge collection and administration in the hands of outside agents or a separate subsidiary company. One reason for this is that most leases provide for the cost of outside agencies to be a service charge item but do not provide for the landlord's own costs to be recoverable. Although some leases specifically allow the landlord to charge a fee (and it is only in these circumstances that he can do so), tenants prefer to see a limit set on such charges; this is quite often expressed as a percentage of service charge expenditure, e.g. ten per cent.

Such arrangements may still not be entirely cost-effective and some landlords have sought to reconcile their desire for direct control with their need for profitability by setting up subsidiary companies, which effectively become managing agents. Since these subsidiary companies are separate legal entities, their costs are recoverable from tenants as a service charge item. This device, however, does not necessarily provide for the true independence which some tenants prefer.

Another method sometimes employed by landlords is to ensure that the leases provide for recovery of both the landlords' charge for management and the surveyors' (as distinct from managing agents') fees. This allows landlords to appoint surveyors to advise and deal with day-to-day management matters, and to recover the costs under the heading of surveyors' fees. While there may be nothing contractually wrong in this method, it does mean, in practice, that there is a double charge. This artificially inflates the total service charge which could be to the landlord's disadvantage in terms of competing for tenants and the rent they would be prepared to pay. This would be of particular relevance at rent review.

Managing Agents

It is generally more cost-effective for landlords to appoint outside agents – usually, but not exclusively, firms of surveyors – to manage the property. Leases must, however, include a specific provision for the cost of managing agents to be recoverable via the service charge. If there is no fee provision, the courts will usually take a strict interpretation of the lease as not entitling the landlord to recover costs.

On the assumption that leases make the appropriate provision, respectable managing agents provide the specialist expertise and resources dedicated to property management, of which service charges form an integral part. These resources usually include, particularly in larger firms of agents who manage several hundred properties, all necessary computer equipment and software. It is the agents who bear the cost of setting up and maintaining such resources and who have the incentive to remain efficient in order to maximise their own profit. A landlord with just a few properties could find it costly to provide the same level of service. Most managing agents will also have access to an array of specialist expertise in areas such as building surveying, investment, agency, property taxation, and planning and development.

It should be noted that managing agents have a duty to act impartially as between landlord and tenant, despite the fact that the landlord makes the apportionment. This is particularly true where agents' fees are payable by the tenants via the service charge. There appears to be a duty of care to both parties and, in any event, the agents must be seen to be acting reasonably at all times. This delicate situation was clearly set down in *Concorde Graphics Ltd. v Andromeda Investments SA* (1982) 265 EG 386.

Management Companies

A distinction should be made between the subsidiary companies of landlords who act as managing agents and those companies set up to manage property on behalf of a group of individual freeholders, each of whom will have a shareholding in the management company. This latter type of company has traditionally managed blocks of flats and other residential properties but, during the 1980s, the principle was widely applied to commercial properties, particularly modern business parks. In this period, stimulated by government policies, the growth of owner-occupation of property by businesses was widespread. Developers were quick to recognise the demand and created, in particular, out-of-town campuses, covering both light industrial and office space (B1), offered to occupiers on either a freehold or a long-leasehold basis. Thus, a small group of buildings – in any shape or form – might be owned by individuals in much the same way as they would own a house, except that there are common areas which require a common maintenance approach and policy. These usually include landscaping and car-parking areas but may also include the exterior fabric of the buildings, common insurance, management, and other services which may be appropriate to the scale and nature of the development. The control of those common services is assigned to a management company, and each of the individual owners has a share in that company, usually proportional to the size of its property or ownership.

Where the properties are owned on a freeholder basis, a separate deed of covenant governs the setting up of the management company, its shareholding and ownership, and its parameters of operation. Where buildings are sold on a long-leasehold basis, the management company may be set up so as to own the freehold interest as well as to provide common services.

The shareholders in the management company may decide to form a committee amongst themselves, to supervise and implement the services and the administration of service charge accounting. However, most realise that this is not particularly cost-effective, especially when they should be engaged in managing their own businesses, and therefore most such management companies are empowered to appoint a managing agent, not only to supervise and administer the services required but also to take on the role of the company secretary to the management company, which should be a registered limited company, and thus deal with all the statutory requirements of that company.

Change of Ownership and Management Agents

With a change of landlord, it is sensible to ensure that each tenant is notified in writing immediately of the different circumstances (it is only compulsory in the case of residential property, by virtue of the Landlord and Tenant Act 1985). Efficiency requires, however, that any new landlord, and also any new managing agents, should take over full records and details of service contracts and other matters relating to the general control of the building and service charge administration. The latter is of particular importance with regard to service charge funds. A prudent purchaser would also have called for copies of the previous years' service charge accounts and summaries and, certainly, full details of the expenditure in the current service charge year. The more thorough the purchaser or his agents can be in the circumstances, the easier the transfer of property will become. All too often these matters are given only cursory consideration, particularly in a rising market, and this can give rise to problems at a later date.

By virtue of s. 142 of the Law of Property Act 1925, the new landlord will be liable to undertake all the landlord's covenants which run with the land, and these normally include service charge covenants. The new landlord will also be able to enforce the tenants' obligations under the leases, including those relating to arrears of service charges arising before the transfer of ownership. Again, however, prudent purchasers should make provision in the conveyance that they will not be liable to inherit such arrears and should insist on the vendors' making up the difference, particularly with regard to the service charge. The likelihood of this succeeding depends, however, on other factors, such as price and market forces.

Expenditure Budget

As a point of good management practice, service charge expenditure budgets should be made, and, where possible, adhered to, for each service charge accounting year. If nothing else, it focuses attention on the tasks ahead. Such budgets should take into account the ordinary routine expenditure likely and reflect an uplift for possible inflation and any predicted items of extraordinary expenditure. These last would include major items of repair, cyclical decoration costs, or plant replacement, for instance.

Such budgets will then form the basis for the advance charges to be demanded from tenants (if allowed under the lease terms) and enable cash flow to be anticipated. Some leases will, in any event, require the provision of budgets to be made available to tenants and may also specify the composition of such budgets, i.e., it may be stipulated that advance charges can be demanded only on the basis of the total of the previous year's ordinary expenditure, plus an uplift for inflation plus extraordinary items. Other leases may allow for the advance charge to be based only on the total of the previous year's expenditure, without any further allowances or uplifts. The rule, as in all matters connected with service charges, is that the lease provisions must be strictly adhered to.

Longer term forward budgeting should also be considered, e.g. three to five years ahead. This will enable large items of expenditure to be identified and appropriate provision made for sinking or reserve funds to avoid oscillation in expenditure from one year to the next. This form of budgeting will be appreciated by tenants who themselves may be required to forecast expenditure several years in advance and whose own financial accounting period may in any event not coincide with the service charge accounting period. This will help landlord and tenant relationships and avoid acrimonious disputes when large items of expenditure arise.

Long-term budgeting can be assisted by planned maintenance surveys in respect of the fabric of the building, by close liaison with any mechanical and electrical contractors and, if necessary, by independent audits by building services' engineers. Certain types of maintenance contracts, if set up correctly at the outset, will avoid excessive variations in expenditure by incorporating plant replacement funds. Lift maintenance contractors, for example, will usually offer what they call a fully comprehensive service, so that, while the annual cost may seem high, it is fully inclusive and there will be no demand for extra payment in the event of major part replacements. Similarly, heating contractors can provide for plant replacement funds or guarantees within their costings. This has the effect of making them responsible for planned maintenance, which in itself will ensure a more efficient service, although this may not enable the landlords to gain the benefit of capital allowance taxation schemes (see Chapter 7 below).

Purchasing Services

Once it has been decided who manages, and the parameters of expenditure have been set to a budget, the purchase of services can proceed, albeit with due regard to common sense, case law, and statutory requirements, as well as the all-important lease provisions.

Most modern leases do not impose too many restrictions on the manner in which services may be obtained by the landlord. However, it obviously makes good sense when setting up service contracts to obtain competitive quotations from reliable suppliers. This does not necessarily mean selecting the cheapest quote each time, but rather one which will provide the best value for money, provided there is not too wide a discrepancy between the chosen contractor's price and the lowest price offered.

Liaison with Tenants

As a matter of good practice, even if not expressly required by leases, prudent landlords and managing agents should liaise with tenants, particularly over major items of expenditure likely to arise in any forthcoming service charge year. It is, of course, not always possible to predict when major items of expenditure will be required, but a well-managed budget will incorporate an element of planning for such large costs. Wherever possible, tenants should be advised of extraordinary items before or at the commencement of the service charge year, to give them the opportunity to comment and plan their own financial contribution. At the same time the landlord may take into account some of the tenants' requirements, especially the major space users, who often have different objectives. Tenants may also wish to check the validity of the landlords' plans; landlords need therefore to be well-prepared and to ensure that any proposed costs are recoverable in accordance with lease provisions.

Consultation which encompasses not only cost but also practicalities and advance warning will be appreciated by tenants and help avoid dispute. In the longer term, this promotes a good landlord and tenant relationship.

Where leases lay down a consultation procedure, it should be strictly adhered to, lest it prejudice the landlord's ability to recover costs, as in *CIN Properties Ltd. v Barclays Bank plc* [1986] 1 EGLR 59.

The Test of Reasonableness

Specific lease terms must be checked and adhered to, but irrespective of such provisions, all costs incurred must be of a reasonable nature and have been commissioned in a reasonable manner. The age-old question of what constitutes 'reasonableness' arises and will vary according to each individual circumstance. There has been case law on the subject, mostly in the context of residential service charges, although the principles can apply to commercial properties. *Finchbourne Ltd. v Rodrigues* [1976] 3 All ER 581 is perhaps a bench-mark case. It was held that not only did the landlords not have the unfettered right to adopt the very highest standard of maintenance and recharge the tenant but it was to be implied that such costs should be 'fair and reasonable'. Subsequent cases, such as *Firstcross Ltd. v Teasdale* (1982) 265 EG 305 and *Gleniffer Finance Corporation Ltd. v Bamar Wood and Products Ltd.* (1978) 37 P & C R 208, have upheld this principle. Conversely, in *Duke of Westminster v*

Guild (1983) 267 EG 762, a commercial property case, the Court of Appeal advised that implied terms should not necessarily be inferred. This was effectively a decision on repairs and it was held that even though the tenant covenanted to pay the cost incurred by the landlord in maintaining a drain under a private road, it was not implied that the landlord had an obligation to maintain the drain. Irrespective of court cases, it would still be good practice to apply the 'fair and reasonable' principle.

Statutory Requirements

The effect of various consumer-orientated Acts of Parliament on the supply of services by a landlord to his tenants is not entirely clear, but practitioners should be aware of the provisions of the Supply of Goods and Services Act 1982, s. 13, which states that the supply will be with reasonable care and skill; s. 14, which states that the supply will be within a reasonable time; and s. 15, which states that the party contracting with the supplier will pay a reasonable charge.

3   DISPUTES

Tenants may raise any of the following queries on the substance of service charges:

1. Does the schedule of services include improvements or enhancement, the cost of which should be funded directly by the landlord?
2. Is what the landlord wishes to include in the charge actually covered by the schedule of services? In *Capital & Counties Freehold Equity Trust Ltd. v BL plc* (1987) 283 EG 563, it was confirmed that a tenant's obligation to pay service charges only arose from costs incurred during the term. The obligation was coterminous with the lease and, accordingly, the tenant could not be required to pay for services implemented and arising after the end of the lease. The landlord had sought to clarify whether the tenant's covenant extended to expenses which *may* have been expended or incurred during the term irrespective of whether they were or not.
3. If it is covered, is it necessary?
4. If it is necessary, is the procedure appropriate? In *Bandar Property Holdings Ltd. v J.S. Darwen (Successors) Ltd.* [1968] 2 All ER 305, it was held that the landlord need not adopt the cheapest method available in supplying the service (as noted earlier). Furthermore, as long as it is reasonable to do so, the landlord can elect to do a permanent job rather than patching up (*Manor House Drive Ltd. v Shahbazian* (1965) 195 EG 283). However, in *Mullaney v Maybourne Grange (Croydon) Management Co Ltd.* [1986] 1 EGLR 70, the court decided that the replacement in a tower block of wooden-framed windows (which required painting every four years) by double-glazed maintenance-free windows,

went beyond what was necessary for the purpose of effecting the repair.

5. If it is appropriate, is the procedure carried out and supervised correctly?
6. If it is correctly carried out, is the cost incurred reasonable? *Finchbourne v Rodrigues* (see above, p. 18) is the authority for the proposition that the landlord can only recover costs from the tenants if such costs are fair and reasonable.
7. Is the certificate of costs required sufficiently explicit?

The landlord's main concern will be the comprehensiveness of the service charges' clause. During the 1970s, lessors gained a commanding position on service charges in both the actual approach to the items to be included and the restrictions on the lessees' opportunity for effective criticism. Some leases contain a final phrase in the schedule of services provided, to include:

> provision of any other service or facility and the making of any other payment which may reasonably be required for the efficient running of the building, the comfort of the lessees and the efficient running of the service areas

together with a clause that the decisions of the lessor's surveyor should be final on any matter, provision, liability, or apportionment.

These 'sweeping-up' provisions do seem to be weighted in favour of the landlord but, on interpretation, the courts do seek to make a literal interpretation. *Mullaney v Maybourne Grange* (see above, p. 20) derived from such a clause. *Jacob Isbicki & Co Ltd. v Goulding & Bird Ltd.* [1989] 1 EGLR 236, sought to determine whether sand-blasting of external walls was within the power of the landlord's repairing obligations even though that type of work was not specifically mentioned. The landlord sought to recover these costs under sweeping-up provisions which stipulated that:

> the landlord may at his reasonable discretion hold add to extend vary or make any alteration in the rendering of the

said services or any of them from time to time if the landlord at his like discretion deems it desirable to do so for more efficient conduct and management of the building.

The court decided that the sand-blasting works were not of the kind originally anticipated in the landlord's repairing obligations and therefore the landlord could not rely on the sweeping-up provisions to recover the costs.

The most important case in this field is perhaps *Concorde Graphics Ltd. v Andromeda Investments SA* 1982 (see above, p. 13). The lease provided for disputes to be settled by the landlord's surveyor, whose decision was to be final and binding on the parties. The tenant queried the inclusion of some items in the account, the cost of other items, including the managing agent's fee, and the basis for apportionment. The court held that while the landlord's managing agents had made a demand for the service charge on the tenant in that capacity, the function of the landlord's surveyor in the clause was essentially arbitral. Although he is the landlord's agent, he must act impartially and hold the balance equally between the landlord and the tenant, notwithstanding that the landlord is his principal and paymaster. His position is no more delicate than the architect required to issue certificates under the standard Royal Institute of British Architects' contract. The managing agents were, however, unable to perform this arbitral function, and, while the court declined to accept the tenant's whole proposal, it did point out that the problem had arisen because the landlord had appointed a firm of surveyors to act as both its surveyors and its managing agents, and that, in the circumstances, the landlord should appoint other surveyors to fulfil the arbitral role.

In the case of a dispute over service charge and in the absence of any expressed provisions in the lease as to procedure for resolution, there are three broad options available:

1. The parties agree to refer the matter to a third party, to act as either an expert or an arbitrator.

2. Where the tenant has failed to pay monies due, as the result of a dispute, the landlord can sue the tenant for the outstanding costs and the tenant can defend its action in court.
3. Either or both parties can apply to the court for a declaration on the validity of the service charge demanded.

Disputes are professionally frustrating, administratively time-consuming, and financially wasteful. While the immediate goal will be to settle the dispute, the management surveyor should search diligently to find the cause, and add it to the list of points to be avoided when setting up the next agreement.

4 ACCOUNTING PROCEDURES

The management of service charges involves substantial accounting input, and, if that is running smoothly, many problems can be obviated. The control of the income and the expenditure is, however, a team effort between property managers (surveyors, administrators) and the accounting function. It can be dangerous to define the accounting role too strictly, as success depends on co-operation between the disciplines.

In general terms, an accounts department (of the landlord or managing agent) will be responsible for (in addition to rent matters) service charge collection and credit control services, management of expenditure, a purchase ledger system, and for quarterly or monthly reporting, and preparation of service charge analyses and service charge accounts, in accordance with the company's policy or, in the case of managing agents, in accordance with the client's instructions, which can vary with each property or client. Although some smaller organisations may still use a manual system of ledger control, computers are generally regarded as an essential and cost-effective tool in the accounting function, particularly given the typical volume of paperwork to be processed.

In the following sections, a summary of the type of procedures which should be implemented, to ensure comprehensive control of the service charge, is given. For convenience and clear identification, it is also desirable that each property is set up as a separate 'cost-centre' for accounting purposes, and the procedure is designed on that basis.

Service Charge Demands

1. Service charge demand records should be created, in conjunction with property managers (assuming all other lease details are also recorded); these records should reflect the variable service charge, ready for inclusion in the demands to tenants (invoices).
2. Draft demand information should be prepared for approval by managers.
3. Formal demands should be prepared and dispatched three to four weeks before the due date.
4. The VAT must be calculated for demands sent to clients (where appropriate) in accordance with the option to tax for VAT purpose(s) (see Chapter 5).
5. Adjusted and *ad hoc* demands must be issued as and when required.
6. Queries from tenants, and clients in the case of managing agents (regarding the service charge demands), must be answered. It may be that all enquiries are routed through an accounts department, but they may need to pass some of them on to the property managers for reply.

Service Charge Collection

1. A daily cash book and a daily computer batch report of service charges collected need to be maintained.
2. Cash receipts should be posted daily to the computer and standing order receipts via the bank should be noted.
3. Service charges received should be monitored and, in the case of managing agents, paid over to the clients as per agreed policy, or, in the case of advance payments, retained to form an expenditure float.
4. Arrears must be monitored, property managers informed regularly, and a credit control system of arrears' letters maintained. A typical procedure would be as follows:

(a) Final reminders are sent out on the due date, usually the quarterday.

(b) One week after the quarterday, an arrears list is produced and sent to managers; at the same time arrears-chasing letters can be produced for dispatch by managers.

(c) One week later, a second, perhaps stronger, letter is sent to tenants.

(d) A further five to eight days later, property managers are requested to authorise further action, such as issuing distress warrants to bailiffs, if appropriate.

(e) Thereafter, a weekly arrears list is circulated to property managers, who ensure that the appropriate action is taken to recover any outstanding arrears.

5. VAT receipted invoices must be produced where appropriate, and dispatched to tenants.

6. Queries from tenants regarding payments must be answered.

Expenditure

1. All invoices received from suppliers must be registered (on computer). In the case of managing agents, invoices must be correctly addressed to their client, care of the managing agents, for VAT purposes. Where invoices are incorrectly addressed, they should be returned to the supplier in the first instance.

2. Once logged into the computer system, invoices should be sent to property managers for approval in accordance with the contract or instructions placed. Utilising a predetermined coding system for each head of expenditure, managers should return approved invoices promptly.

The coding of invoices is a simple accounting procedure – each type of expenditure is allocated as a numerical identity, recognisable by the computer. An example would be: external repairs – code 004; electricity – code 002. These are effectively heads of expenditure and the fewer there are, the easier the service charge accounts to prepare at the year end.

3. Approved and coded invoices should be posted on the computer.

4. Client cash balances and payment to suppliers, if funds are adequate, must be monitored in good time to avoid unnecessary reminders from suppliers.

5. Where, in the case of managing agents, funds are inadequate either the client must be informed direct or the managers must be reported to for further instructions. Cash flow management, particularly where quarterly advance payments are made, is important as, although income will be regular, expenditure will not be.

6. Suppliers' invoices have to be monitored, their queries answered and managers prompted, if invoices have not yet been approved or coded.

Quarterly Accounting (or Monthly if appropriate)

At predetermined times, and as soon as possible after each quarterly or monthly accounting date, the accounts team should prepare draft management accounts, for review by property managers. The account also needs to be checked at this point for VAT implications and, where appropriate, any required Inland Revenue deduction for 'off-shore' clients made.

The account should adequately present all income and expenditure, setting out the balance in hand for items of expenditure, either the service charge held or, in the case of managing agents, the float provided by the client. Once the account is satisfactorily reconciled, full copies should be prepared and dispatched with the original suppliers' invoices (in the case of managing agents) attached. If necessary, the account may be accompanied by the balance of any payments to be made to the client, or may request payment from the client to maintain the expenditure float.

Each quarterly account also represents an opportune time to review expenditure. If it is exceeding budget, it may be necessary to request additional funding, as the tenants cannot usually be asked to increase interim payments during the accounting year, except by agreement.

After the quarterly account has been dispatched, arrears must continue to be monitored and the income must be passed on to the client (by managing agents, if appropriate).

General Accounting Matters

The accounts team also needs to administer and set up the systems to deal with the following areas:

1. correspondence, registration of all bank accounts, including individual client accounts, general client accounts, and deposit and interest-bearing accounts;
2. cash books with bank statement on computer, and a petty cash system;
3. clients' and all cost-centre balances, to ensure that no account is overdrawn;
4. compliance with the RICS Client Account Rules* at all times, in the case of managing agents which are chartered surveyors;
5. compliance with the VAT Regulations;
6. compliance with Inland Revenue Regulations regarding clients, not domiciled in the UK for tax purposes;
7. PAYE services in respect of staff employed in the buildings. In the case of managing agents, this service may be handled on behalf of clients;

* For those practitioners who are members of the Royal Institution of Chartered Surveyors, certain rules and regulations govern the handling of clients' money, over and above any requirements embodied in the Estate Agents Act 1977. All practitioners should ensure that they, and particularly their accountants, are familiar with these requirements.

8. a daily and weekly security routine to avoid loss of computer data;
9. preparation, in conjunction with the property managers, of an annual report on the service charge accounts, both to explain variations to the budget and to forecast expenditure.

Service Charge Year End Accounts

At the end of the accounting year, certified statements of expenditure have to be produced for the tenant. With each property set up as an individual cost centre, all expenditure can be recorded against that cost centre in accordance with the coding system for individual items of expenditure (see above, p. 26). The accounts team, who should have regularly monitored expenditure and reported to managers, must add up the service charges at year end (this will be a statement of the actual service charge expenditure), making the necessary adjustments for accounting or other areas, and compare totals with the original budget. Thereafter, with the agreement of property managers, expenditure has to be apportioned in accordance with tenants' leases.

Thus, end-of-year responsibilities consist in:

1. Reconciling the service charge balance with the (client's) balance, taking account of service charge monies received in advance and the expenditure for the year.
2. Receiving from the property managers the service charge budget for the coming year and liaising with them where appropriate.
3. Sending the final account (to the client) for approval or comment, if appropriate; if required by the lease it must also be submitted for audit. Once it is audited or approved by the client, certified copies must be issued to tenants with demands for balances due where actual expenditure exceeds advance payments or with credits as appropriate.

4. Finally, computer records must be updated with service charge adjustment demands, and revised demands made where required in accordance with the forthcoming year's budget.

Certification of Accounts

Modern service charge provisions require the landlord or managing agents to prepare the annual accounts, and that they should be certified as being correct by either the landlord's agents or an accountant. Occasionally, leases stipulate that the accounts are to be prepared and certified by the landlord's surveyor which, following *Finchbourne Ltd. v Rodrigues* (see above, p. 18), will usually be independent of the landlord, and probably not the managing agents.

As stressed, the lease provisions must be followed precisely, and the expenditure that is certified can only include what is specified within the lease. (See Appendix 1 for a sample service charge certificate.)

Many service charge provisions stipulate that the certificate is totally binding upon the tenant and conclusive. The correct view is that it can only be conclusive on the question of fact, which cannot, of course, obviate the jurisdiction of the court on a question of law. Accordingly, while it may be a question of *fact* that expenditure was made on an individual item, whether that item of expenditure should have been included at all is another matter over which the courts would have jurisdiction.

Inspection of Invoices

Leases do not often make specific provision for tenants to be able to inspect invoices. Service charge certificates by their very nature are a summary of that expenditure under itemised headings. Tenants may feel that they wish to check that the expenditure has actually been incurred, and sight of the relevant invoices would usually be sufficient evidence in this respect. Even if leases do not give the tenant such a right, it is good practice to allow tenants to inspect; landlords avoiding this procedure may only generate distrust and threaten a breakdown in the landlord–tenant relationship. In any case, were matters to escalate into a full-scale dispute, the rules of discovery would allow the tenant to see the invoices anyway.

5 VAT

The question of VAT for service charges is at present governed by the Finance Act 1989 which, with effect from August of that year, introduced a new set of rules and regulations pertaining to property generally. While, doubtless, practitioners are already familiar with the effects on developments and rents, they may not be aware that they also apply to service charges.

The regime now applying to service charges is, in many ways, a substantial improvement on the law prior to August 1989, which was beset by a web of interpretation and application by landlords, tenants, and HM Customs and Excise. The current position is set out below, preceded by a brief summary of the principles of Value Added Tax.

First, it is necessary to distinguish between exempt and taxable goods and services. Exempt items are outside VAT. Taxable items may either be standard rated (currently 17.5 per cent) or zero rated.

Second, VAT is really two taxes – *input* tax (on goods and services supplied to you or a property, i.e. tax you have paid), which is effectively a credit, and *output* tax (on goods and services you supply, i.e. tax you have collected), a debit. Accountability to Customs and Excise is measured by the excess of output tax over input tax.

The application of VAT to service charges depends on whether the landlord or owner has 'elected to waive exemptions to tax', i.e. has chosen the option to tax. Whether or not a landlord opts to tax a building for VAT purposes will be largely governed by non-service charge considerations in the short term. For example, if the landlord has a new development with VAT to be paid on construction, or a forthcoming refurbishment with

VAT on costs, or has purchased a building subject to VAT, it will wish to recover or offset the input tax against output tax.

It is worth remembering that once an election to tax has been made, it is irrevocable whilst the property remains in the ownership of the landlord who has made that option. A new owner could, however, alter the election; this is unlikely in practice, if only because the new owner will have had to pay VAT on the purchase price.

Options to tax can, on the other hand, be made on a building-by-building basis, i.e. they are not confined to whole portfolios. This may be particularly useful where very large service charge expenditure is expected and it is desirable to soften the impact on tenant costs in any one building. However, the administrative burden of VAT must not be under-estimated.

It may be noted that a report published in 1989 (Knight Frank & Rutley Research, 'The Impact of VAT on the Commercial Property Market') suggests that an option to tax is only of small financial advantage to standard-rated occupiers paying a service charge. It commented that service charges generally (where they apply) usually amount to some 5-10 per cent of rents, and thus the recovery of the VAT element within service charges is only a peripheral benefit, representing less that a one per cent reduction in occupancy costs.

No Option to Tax

In these circumstances, all service charge costs will be apportioned and recharged to tenants on a gross basis, i.e. inclusive of all VAT. The service charge budgets will need to take into account the gross basis, and the advance payments demanded of tenants and the reconciling balances must all be shown on the gross basis. There should be no reference to VAT as a separate item at all.

Clearly, goods and services supplied to the building will be subject to VAT, and that VAT must be paid. If the landlord does not opt to tax the building, however, this is irrelevant for service charge purposes; it is the gross figure which is important.

Under this option, the tenants who are registered for VAT are unable to seek VAT invoices from the landlords and are thus unable to recover their input tax. Some tenants find it difficult to understand this, because if they had purchased the goods and services direct themselves, they would have been able to recover the input tax. However, once it is explained, most tenants accept the situation – indeed they have little choice.

Option to Tax

Where the landlord has opted to tax, service charge demands must be expressed as the net cost figure plus VAT, i.e. in the normal invoicing manner. From the tenants' point of view, especially if they are registered for VAT themselves, this is a straightforward transaction: they pay an invoice which is subject to VAT, and that VAT is clearly identified, which, in turn, means that they can treat it as an output tax to be offset against their own input tax in their accountability to Customs and Excise.

From the point of view of the landlords and of their managing agents, however, there are further administrative considerations. Customs and Excise Notice 742B (Property Ownership, January 1990) specifies that the tax point, the date at which VAT on service charges should be accounted for, is either when payment is received or when a tax invoice is issued, whichever happens first. This rule, however, does not take into account the practicalities, i.e. that most rent and service charge demands are issued some weeks in advance of the due date which is usually a quarterday. The quarterday itself is not therefore necessarily the tax point: payment is quite often received after the rent quarterday.

Accordingly, in order to avoid the situation where a rent demand could be treated as a tax invoice, several weeks in advance of the date at which the payment is actually required under the terms of the lease, most landlords and their agents have taken to issuing separate demands and tax invoices. Thus, in the usual manner, prior to the quarterday, the landlord will issue a demand, which quite clearly states that it is not a tax invoice, on a fully inclusive, i.e. gross, basis. When payment is received from a tenant, a tax invoice is issued by the landlord, confirming that the VAT is paid. This is obviously an administrative burden, and its impact upon the cost of managing elected properties will need to be taken into account by landlords and their agents.

On the other hand, the effect of these costs may be mitigated by the avoidance of a VAT liability arising at a tax point falling considerably earlier than the date at which the tenant actually pays. This is particularly important in the case of default or late payment. Furthermore, the date at which the landlord has to account to Customs and Excise may be several weeks past the date of receipt of income from tenants and thus the landlord has the benefit of holding the money in that intervening period.

Subsequent administration of VAT is largely one of accounting practice, in which most companies are well versed. There are, nevertheless, some specific procedures for managing agents to consider.

Property managers need to ascertain from their client: the date upon which registration commenced; the client's VAT reference number; in which quarters the client's VAT declaration is due; and the client's preference for VAT invoices to be on the demand or on the receipts basis. They need to inform the accounts department of the above and update both clients' and tenants' records on the computer to indicate VAT status.

The accounts department needs to check that the following are correctly entered on the computer: the client's VAT registration number, the tax point (whether on demand or receipt), the client's and tenant's VAT status. It must record the client's quarterly VAT details to ensure that management accounts are sent out at the appropriate time. It must send a copy of every VAT invoice to the client immediately upon production, and hold a copy of every VAT invoice in the accounts file for later inspection by the VAT officer. Where VAT invoices are required, it must ensure that these are produced promptly after payment is received. Finally, quarterly accounts must be checked to ensure that the VAT analysis is correct, and any balancing sums due to or from the client must be calculated and explained on the quarterly management account.

Further information on VAT as it affects service charges, and indeed property ownership in general, can be found in Customs and Excise Notice 742B (January 1990); for managing agents, Notice 700, the VAT Guide Section IX, explains how agents should account for VAT and what to do.

6 ARREARS

The treatment of arrears depends on whether or not the service charge is reserved as additional rent; this is quite possible in the case of rack rented leases, but less likely under long leases and deeds of covenant. From a practical point of view, it is generally preferable to have the service charge reserved as rent, as the remedies for recovery are both wider and easier to implement.

Service Charge as Rent

Where an advance payment falls into arrears, the landlord has the same opportunity to recover it as he would actual rent. Thus, the simple and cost-effective remedy of distress is often employed. Distress is the right of the landlord to take the goods of the tenant on the demised premises in order to obtain payment for rent due and in arrears. The threat of this is usually enough to persuade defaulting tenants to settle, for fear of having goods removed from their premises.

No legal process is necessary for the application of distress, but it must be applied correctly. The law is complicated with statutes and cases going back three hundred years; and it is therefore recommended that implementation should be by an experienced certificated bailiff (for further discussion, see Adkins, *Landlord and Tenant*, Estates Gazette, 1982). Used judiciously, this method is relatively cheap, although the defaulting tenant is bound only to pay the bailiff's statutory levy, and there is usually an additional bailiff's cost to be borne by the landlord. These costs can be ascertained and agreed in advance. It is also advisable to warn tenants, in writing and in advance, of the intended use of this method if payment is not made.

Even greater care needs to be exercised in recovering arrears by this method where a balancing sum is due. The amount must have been finally ascertained in accordance with the lease terms, since, until it has, payment will not actually become due and thus the arrears, as such, will not be validated.

The use of distress is not recommended as a form of recovery where there is a genuine dispute. In these cases, it is best to proceed through one of the other forms of action (see below). It should be remembered, too, that the remedy of distress may result, if payment is not forthcoming, in forfeiture of the lease by peaceable re-entry (see below). Distress should, therefore, always be considered in conjunction with the desirability of having vacant possession of the premises. In some cases, particularly in a poor letting market, it may not be an advantage to have possession, and, thus, pursuing the arrears through the courts may be more appropriate, whilst also keeping the tenancy alive. This will have to be weighed against the strength of the tenant's covenant – clearly, if the tenant is in financial difficulties, there may not be much to be gained by prolonging the tenancy.

Court Action

Alternatively, where the arrears are valid (i.e. there is no dispute), and the service charge is recoverable as additional rent, a rent action can be brought in the county court, provided the court has jurisdiction – this is determined by the prescribed rateable value of the premises and the level of the arrears. If the relevant county court does not have jurisdiction, application can be made to the High Court.

Service Charge not Recoverable as Rent

In these circumstances, where there is no dispute, the arrears are treated in the same way as any other debt where there has been a default in payment. Subject to the jurisdiction limits of the courts, a summons for non-payment is issued. In such circumstances, property managers should have recourse to solicitors, to ensure the matter is correctly handled.

Forfeiture of Leases

Most leases provide for forfeiture in breach of covenant, and non-payment of service charges would obviously be thus categorised. Leases also generally provide for the procedure under section 146 of the Law of Property Act 1925 to be followed. This action can be taken irrespective of whether the service charge is reserved as rent or not.

The relevant section 146 notice must be served specifying the breach of covenant and giving a reasonable time to pay. That period may depend on what action and application for payment has already been made. Some leases specify a time period, fourteen or twenty-one days after the due date, say, before which any failure to pay cannot be treated as an arrear.

Where a tenant still fails to pay and the matter proceeds to a court hearing, the tenant can apply for, and the court can grant, relief against forfeiture, but, in the case of a clear arrear situation, relief is likely to be granted only on condition that the tenant makes payment within a specified time. The disadvantage of forfeiture proceedings is the time it takes to achieve a result.

Some leases also give landlords the right to forfeit the lease without the need to serve a section 146 notice, merely by peaceably re-entering the premises. The House of Lords' decision in *Billson v Residential Apartments Ltd.* [1992] 01 EG 91 came as a timely reminder as to what actually constitutes peaceable re-entry. It is recommended that landlords serve notice of their intention to re-enter, and peaceable re-entry can then take place provided that there is no one on the premises to oppose that re-entry. Entry can be peaceable by breaking the lock or gaining access via a window, provided that the person effecting the entry knows that there is no one on the premises to oppose it. The question of there being no opposition present is covered under the Criminal Law Act 1977.

It was established in the last century (*Quilter v Mapleson* (1882) 9 QBD 672 and *Rogers v Wrights* [1892] 2 CH 170) that once re-entry has been effected the court has no jurisdiction, i.e. the tenant is not then able to apply to the court for relief. Bearing this in mind, a tenant receiving a notice from a landlord may wish to consider applying to the court to obtain relief against forfeiture. The tenant will have to act quickly in order to activate the court's power and give time for the application to be processed so as to prevent the landlord from determining the lease by peaceable re-entry. Each case will, of course, rest on individual circumstances.

Assignments, Surrenders, and Arrears

It is essential that all arrears are settled as a pre-condition of any consent to an assignment or a surrender of the lessee's interest. Apart from making commercial sense, it should be noted that a new tenant, as assignee, is probably not liable to pay arrears which arise before the date of the assignment. Such arrears could be deemed to be 'once and for all' breaches for which a new tenant was not responsible. Of course, an assignee could be liable to pay where it has specifically undertaken to pay past rents and service charges, but the best option for the landlord would still be to require clearance of arrears at the outset.

Practitioners need to consider, too, what happens to reconciling balances at the end of the service charge year – those charges relate, of course, to the whole of the service charge year, during which there would have been two separate tenants. It may be argued that the reconciling balance only becomes due at a specific time and is not an arrear until it is declared due, and the current tenant therefore becomes liable. It should not be forgotten, however, that the original tenant, the assignor, can still be sued for non-payment of the service charge arrears even after the date of the assignment (see below).

Furthermore, where a lease requires the landlord's consent to an assignment, as is usual in most modern forms of lease, the landlord would be entitled to withhold consent if there were substantial arrears and could not be held as being unreasonable in doing so. Landlords would also wish to consider the position of arrears in negotiating or agreeing to surrenders.

Previous Tenants and Sureties

Where a tenant defaults in service charge payments, the lease should be examined carefully to ascertain the obligations of the sureties or guarantors in such circumstances. The procedures for recovering arrears from such sureties should then be followed through, if necessary, to court action.

By virtue of the privity of contract which exists between a landlord and the original tenant, that original tenant can still be sued for arrears arising under the lease despite the fact that it may have assigned its interest several years previously, and that there may have been intervening assignments. Licences to assign and deeds of consent usually require assignees to give a direct covenant to pay rents, service charges, and other sums due, throughout the residue of the lease. This being the case – and licences should, of course, be checked in this respect – the previous assignees could also become liable for payment of arrears.

7   RESERVE OR SINKING FUNDS AND TAX

Reserve or sinking funds and tax may appear to be separate issues, but they are in fact inextricably linked and are thus treated together here.

Reserve or Sinking Funds

In order that tenants' contributions for major items of expenditure, e.g. extensive redecoration, plant and machinery replacement, may be spread evenly and fairly over the life of the building, the creation of a sinking or reserve fund appears to be an ideal solution, avoiding wide oscillations in yearly expenditure. The funds can be held with independent trustees to satisfy tenants that the accounting is being properly done. Yet, although the idea seems simple, the practicalities are not and usually cause more problems than they solve.

Some major tenants regard reserve funds as unnecessary, since they are prepared to meet a variable service charge on the basis of cost incurred, knowing that overall they will make the same contribution. Smaller businesses usually find such funds attractive.

Some landlords have concluded that the administration and tax implications are unacceptable or, at least, not worth pursuing. While the tenants will be anxious to ensure that contributions to the fund are treated as an expense in the same way as rent, the landlord will want to avoid taxation of the fund and its accumulation of interest. Unfortunately neither position is tenable.

The foregoing assumes that a given lease provides for the creation of a reserve or sinking fund in the first place. But, even if the lease does not make such a provision, it may be possible to draw up agreements between landlords and tenants anyway, though this is quite often fraught with difficulties, and there will be additional costs in negotiating and concluding an agreement which is in a form acceptable to all parties.

Sinking Funds and the Tax System

It seems that the tax system has no fixed provisions for a sinking fund, i.e. a fund for future expenditure, with its estimates and future provisions. The treatment of such sums varies from one tax inspector to another and there is arguably a need for some legislation or general clarification to encourage the operation of such funds.

At present there are four main ways in which sinking funds are set up for tax purposes.

The Fund as Additional Rent

Where the contribution is expressed as additional rent in the lease, the landlord is likely to be taxed on receipt of that contribution – as normal income – but will only be able to set the expenses (to be incurred in the future) against income during the accounting period in which the expenditure actually arises. In other words, the landlord cannot claim a tax deduction for funds to meet future liabilities. The landlord may, therefore, carry a tax burden for several years before any expenditure is initiated on which relief can be obtained. This may not be attractive to a landlord and poses the question as to what happens if a sale takes place in the interim. It may be possible, depending on the circumstances at the time, to require a purchaser to set up a similar fund, at the existing level, with the vendor reducing the sale price of the investment by an equivalent amount.

The Fund Charged under Covenant

If the fund is not referred to as additional rent, it is charged under a covenant and, strictly speaking, contributions should be taxed as a trade, under Schedule D, which may be less attractive than Schedule A, which will normally be the scale applied when the contribution is made as additional rent.

A Trust Fund

A trust fund could be established, and tenants required to make payments in accordance with the lease terms. The capital received would not be taxed, but the income resulting from the investment would be liable to taxation. The tenants may have difficulty in obtaining relief from tax for their contributions, since it can be argued that the money has not been spent but rather has been allocated for a purpose. On the other hand, the trust will offer the advantage of protecting the tenant from the landlord's dealings in the fund or insolvency. A fund in the form of a trust may well be subject to inheritance tax, although it is doubtful whether it was intended that sinking funds should fall within the parameters of this regime.

Company with 'Mutual Trading' Status

A separate management company could be established by the tenants, or possibly even the landlord, and could be treated favourably for tax purposes were it to have 'mutual trading' status. Schemes of this kind have been established in residential blocks, but they appear to have been less favoured for commercial properties.

The Sinking Fund and the Tenant

While, from the landlord's point of view, the treatment of reserve or sinking funds for taxation purposes appears to be a pitfall to their establishment, from a tenant's point of view, the desirability of making a contribution poses a series of questions as to:

1. the tenant's faith in the landlord or the managing agents,
2. where the funds will be held and who will benefit from any interest accruing,
3. what will happen if it assigns the lease,
4. what will happen to the proportionate balance of the fund at the end of the lease, and
5. how it can be assured that the level of contributions is correct.

These and many other issues can be settled on a practical basis: if the landlord is seeking to establish a fund (although more often than not it is the tenants who request such an arrangement), then the landlord or the managing agents should be as open as possible in order to engender support for it. The main problem is, of course, the unpredictability of future expenditure and changes in circumstances; there is probably little point in establishing a reserve fund if the landlord thinks it is likely to sell the building within the next few years or if several leases are going to come to an end relatively shortly.

Where a long-term view can be taken, the establishment of a fund could be appropriate, but it is probably still worthwhile ensuring, where possible, that funds are expended before the end of leases and that a separate reserve fund is set up thereafter with tenants who renew their leases.

It is also worth stressing that where leases allow for the creation of reserve or sinking funds for specific purposes only, expenditure should only be on those specified items, unless the tenants are prepared to agree otherwise and vary the lease accordingly.

Other Tax Implications for Service Charges – Capital
Allowances

Expenditure on capital items cannot generally be set against tax
in the same way as maintenance and repairs in general. However,
annual allowances can be claimed in respect of expenditure on
qualifying items of plant, such as boilers, lifts, air-conditioning
equipment, but only when they are first installed. The current
annual allowance is twenty-five per cent of cost and, while this
is a useful benefit to a landlord, a tenant cannot gain any
advantage unless the landlord agrees to assign its allowance to
the tenant. This would be entirely at the landlord's discretion,
and probably would only work effectively where a single tenant
was involved, and it had been agreed as part of the lease
negotiations.

Very few leases, however, make any provision for the
repayment or deduction of capital allowances in favour of
tenants in the context of service charges. In any event, tenants
may take the view that they would rather treat the whole service
charge as maintenance, and repairs as an occupying expense or,
indeed, further rent, in which case that expenditure can be set
against revenue and, ultimately, a hundred per cent tax relief
obtained.

Renewals and replacements fall outside the scope of capital
allowances, and, as a 'revenue' item, again, mean a hundred per
cent relief for the tenant.

It is worth mentioning that some items of expenditure which
may be classified as 'repairs' under lease provisions may be
treated by the Inland Revenue as 'improvements' and not as
either plant or normal maintenance and running expenses; items
which constitute improvements for tax purposes are not deduct-
ible expenses. In such circumstances the tenant will not obtain
any tax relief, but the landlord has the option of avoiding taxation
by classifying such expenditure as a revenue item.

Construction Industry Tax Deduction Scheme

The Inland Revenue will sometimes require the landlord or managing agents to account for tax which would otherwise be payable by suppliers of business services. This system is rooted in the construction industry, and the same rules apply to building owners who are also registered as developers or building contractors. Any supplier of services (of a construction nature) to the owner is in this case treated as a subcontractor. The Inland Revenue's rules require the main contractor (i.e. the building owner) to withhold income tax from payments to those subcontractors, i.e. the payment of an invoice must be net of income tax at the prevailing rate.

This is clearly an administrative burden to landlords and managing agents but can be avoided by ensuring that the supplier of the service (the subcontractor) is exempted from the scheme by the Inland Revenue. Only suppliers who can produce a Construction Industry Tax Deduction (CITD) exemption certificate should therefore be used. Landlords and managing agents could be liable for the payment of tax unless they are able to produce evidence of such a certificate. Where a certificated contractor is used, the administrative burden is swept away since that contractor is effectively recognised by the Inland Revenue as being able to account for its taxation direct.

8 COMPUTERS

If the management of property and service charges are famous for one thing, it must be the volume of paper that requires processing. A management department of average size will generate literally thousands of invoices and payment transactions. An efficient computerised system is, therefore, almost essential for all but the very smallest portfolios.

This chapter will offer a brief guide to what a practitioner will need from a system. Each organisation will have different requirements for the computer, varying resources to manage the system, and varying abilities to meet the cost of purchase; consequently, no cast-iron guide to the best type of system can be given. However, whatever system is used, it is essential to remember that the quality of information retrieved from a computer is only as good as the data put into it at the outset.

Application

The purpose of a new or replacement computer should be considered before a purchase is made. Is the aim to acquire a pure accounting system or is there to be a greater emphasis on property information? Possibly, it is a combination of both, but the answer must also depend on the resources available to manage a system continually.

The larger property-owning institutions, who take a long-term view of ownership, quite often favour a system which gives instant access to full details of the property itself, but in which the accounting aspects are secondary, since the accounting process is usually handled by outside agents. Managing agents, on the other hand, probably require an accounts-driven system, with the property information occupying a slightly less important place.

It is a question of evaluating which side of the property management business will derive most benefit, on a cost-effective basis, from a computer. Contrary to popular belief, computers do not necessarily save on human resources and costs, at least not initially, but come into their own when economies of scale can be achieved. A computer should be viewed as a tool, which enables all information on a property to be mixed and matched and presented in a centralised form, eliminating many of the labour-intensive and boring tasks of calculation and data gathering.

Several systems exist for the property world: some are orientated towards owners, while others are more suitable for managing agents. Some systems offer program-writing facilities for those who have the resources. It is inadvisable to rely on what computer companies say. It is best to try to see the systems in operation in organisations of a similar size, preferably where they have been in use for at least two years.

Software companies offer 'off the shelf' systems which have been tested and refined over the years. Computer hardware companies are more interested in selling the actual equipment, often proposing tailor-made software. It is important to be wary of the computer industry's jargon and, given that the property industry is still not very computer literate, to ensure that the system to be purchased can be easily understood and used by all levels of staff – if it is not, it will not be used efficiently or effectively.

Accounting

For accounting purposes, computers should be able to deal with:

1. Income collection – production of rent and service charge demands, as well as invoices for insurance and other costs;

recording the income, producing reminders, and posting income to different clients and cost centres.

2. A cash book.
3. A purchase ledger, processing all invoices through authorisation, payment, and posting to different clients, cost centres, and contractors; and automated cheque printing.
4. The production of income and expenditure statements on a cost-centre and client basis.
5. Audit trails.
6. The generation of debt collection letters.
7. Fully automated service charge account calculation at the year end, including apportionment to individual tenants and production of supporting paperwork. Since each property normally has its own characteristics and methods of apportionment, the ideal system will be able to deal with all the variations. In practice, it is unlikely that any single system will be able to.
8. The production of draft reports on all aspects of accounting, e.g. rent demands. This should allow the items which are to be shown on a demand to be checked prior to the printing of the actual demand, thereby saving time on making more complicated amendments through the computer at a later date. The same applies to service charge accounts and other aspects.
9. The calculation of back rent due on review and demanding the same.
10. The calculation of interest on late payment of rent and service charges and demanding the same.

Management

Property management information – basic lease and property details – is essential for the accounting function to work properly and, thus, an accounts-only package is inappropriate; a property management account system is required. The data necessary to set out and combine the two systems, from a management viewpoint, includes:

1. All tenancy details, the lease commencement and expiry dates, rent review dates, service charge periods, and demand dates. All tenant details – such as names and addresses, changes of tenants on assignment, expiries and surrenders.
2. Actual rent and service charges payable.
3. All basic property details, which may include the service charge parameter dates and the creation of different schedules for this purpose. For example, in a multi-tenanted office building the leases will probably provide that all tenants contribute to all the services, but if, however, there are shops on the ground floor, these will probably not be required to contribute to the maintenance of, for instance, the lift, or air-conditioning to the upper-floor offices. The service charge would therefore have to be set up in two different schedules to cater for these variations, i.e. one to cover all the expenses payable by the offices and another to cater for the expenses payable only by the shops. Without this information, the automatic calculation of the service charge account will not be possible.
4. Details of the clients, name and address, and banking details.
5. Lease interest charging provisions.
6. Rent and service charge stop facilities, so that these items, together with the insurance, cannot be charged beyond the expiry dates without express authorisation. This facility should also enable payments offered by tenants to be blocked when a stop notice is enforced, so that the cash cannot be posted against the tenant, or act as a trigger to reject the payment proffered by the tenant, if necessary.
7. An advance diary date reminder system for lease renewal, rent reviews, decoration dates, and other appropriate tenancy details.

Building Management

Individual computerised building management systems can be useful, in more sophisticated buildings, in assisting in the apportionment of cost and service charges. Such systems can, for instance, identify when any particular tenant has been using the building out of normal hours, and, thus, any additional costs, e.g. for heat or security, can be charged directly to the tenant concerned, thereby reducing the service charge overall. Similarly, heating and air-conditioning can be controlled by maintenance contractors, via a computerised link, to ensure that the plant is delivering heat to the building at the right time and at the right level, thus minimising energy consumption and saving service charge costs in this direction.

There is a wide range of potential applications in this respect, although these are not always cost-effective, and not therefore necessarily of any real benefit.

APPENDIX 1   SAMPLE SERVICE CHARGE STATEMENT

LANDLORD

BUILDING

STATEMENT OF SERVICE CHARGE EXPENDITURE
FOR THE TWELVE-MONTH PERIOD 01.01.199- TO 31.12.199-

£

Insurance premiums	6,606.56
Maintenance, repair, redecoration, and renewal	12,661.14
Internal redecoration programme & carpets	28,658.00
Entrance lobby redecoration	23,772.78
Electricity to office suites	separate charge
Electricity for air conditioning and heating plant, lifts, common services, and lighting	36,578.77
Fuel and maintenance contract – air conditioning and heating	61,140.84
Cleaning of windows and common parts	11,573.66
Landscape maintenance and planters	2,256.27
Lift maintenance and telephones	9,167.91
Water service charge	6,685.37
Provision, maintenance, and renewal of fire systems and equipment	2,249.26
Staff costs	26,139.83
Staff accommodation and telephone	3,405.77
Security	3,500.81
Provision, maintenance, and renewal of signs	673.90
Provision of toilet requisites and towels	7,631.34
Trade refuse disposal	1,973.40
Car park barrier maintenance	378.73
Management	15,525.00
TOTAL	£260,579.34

We hereby certify that from the information available to us, the above statement of the service charge expenditure records the true cost to the landlord of providing the services to the premises for the period to which it relates, in accordance with the terms of the lease.

. . . .
MANAGING AGENT OR ACCOUNTANT

APPENDIX 2   SAMPLE APPORTIONMENT SCHEDULE BY WAY OF CLEAR EXPLANATION TO TENANT

LANDLORD

BUILDING

SERVICE CHARGE EXPENDITURE
FOR THE PERIOD 01.01.1993 TO 31.12.1993

TENANT: Name

FLOOR: Demise Description

TOTAL SERVICE CHARGE EXPENDITURE   £260,579.34

Your proportion under the terms of your lease: 6.70%

i.e. £260,579.34 x 6.7%   £17,458.82

Less advance payments
25.12.1992	£3,000.00
25.03.1993	£4,410.00
24.06.1993	£3,705.00
29.09.1993	£3,705.00

£14,820.00

BALANCE DUE   £2,638.82

APPENDIX 3   SAMPLE REPORT TO ACCOMPANY SERVICE CHARGE ACCOUNTS

This can be particularly helpful and relevant to large accounts and is often appreciated by tenants.

LANDLORD

BUILDING

REPORT ON 1990 SERVICE CHARGE ACCOUNTS

1.0 At £   , total service charge expenditure equates to £   per sq. ft. (Excluding the extraordinary external repair works of £   , 1990 expenditure shows a decrease of   % on 1989 expenditure.)

2.0 The increase in expenditure during the 1990 service charge year has arisen from several one-off items, detailed below and previously advised to tenants.

2.1 As advised in our 1989 Report, external repair works have been carried out to the building and 66% of the anticipated cost of these works was included in the 1990 service charge year with the balance of costs to be included in the 1991 service charge year.

2.2 Redecoration and re-carpeting works to common areas have been carried out during the year, again, as notified to tenants in the 1989 Report.

2.3 The rolling programme to replace perished rubber window seals has continued during the year with £   being expended during 1990.

2.4 As part of the redecoration of the entrance lobby, the existing furnishings and fittings have been refurbished at a total cost of £   .

3.0 As noted in previous reports, direct comparisons on a year-to-year basis are not possible, as invoices received each year do not necessarily synchronise with the service charge year end. However, we set out below the following comments for your information.

3.1 INSURANCE PREMIUM
Insurance premiums are lower than in the previous year accounts as the Landlord has been able to negotiate a better rate with the insurers on the current sums insured.

3.2 MAINTENANCE, REPAIR, AND RENEWAL
As stated above, this sum includes a one-off figure for replacement of rubber seals and draught proofing in the building. Electrical works have also been required to replace various fluorescent fittings in the common parts in the sum of £   .

3.3 FIRE ALARM SYSTEM AND EQUIPMENT
In view of the other large items of expenditure during the year, we have

not carried out any works to the existing fire alarm system which continues to operate satisfactorily. Some replacement will be necessary but we hope to be able to postpone this until the 1992 service charge year.

3.4 STAFF COSTS

As noted in previous reports, staff costs include overtime payments made to the building manager to enable him to carry out necessary maintenance works out of lease hours. This obviously reduces the cost shown under the category of repairs and maintenance.

3.5 STAFF ACCOMMODATION AND TELEPHONE

The large increase under this head of expenditure is due to necessary repairs and refurbishment being required to the kitchen in the building manager's flat.

3.6 CAR PARK AND BARRIER MAINTENANCE

There have been a number of problems experienced with the car park barrier and this increase in costs reflects additional call-out charges.

3.7 M & E EXPENDITURE

This sum was for an inspection and report of M & E services at this property by consulting engineers.

4.0 1991 EXPENDITURE

As previously advised, the balance of external repair works has been included in the 1991 service charge year. The final contract price was lower than originally anticipated and, therefore, tenants' balancing payments are considerably less that the 34 per cent of the total anticipated cost advised in last year's report.

A major cost to be incurred during this year is the start of a repair and renewal programme to toilets throughout the building. Many of the fittings are original (twenty years old) and are in need of renewal. Prior to the commencement of works, tenants will be given further details of the scheme and programme envisaged. The intention is to spread the cost over two years.

In addition to an inflationary uplift on the 1990-1 expenditure, tenants will be aware of the VAT increase, which will raise all service costs by a further 2.5 per cent, and the new advance service charge payments reflect these additional items and costs.

APPENDIX 4   SAMPLE LEASE PROVISIONS GOVERNING CERTIFICATE PROCEDURE

1. The amount of the Service Charge shall be ascertained and certified annually by a certificate (hereinafter called 'the Certificate') signed by the Landlord's Accountant being a qualified Chartered or Certified Accountant as soon after the end of the Landlord's financial year as may be practicable and shall relate to such year in manner hereinafter mentioned

2. The expression 'the Landlord's financial year' shall mean the period from the 1st day of January to the 31st day of December (both days inclusive) or such other annual period as the Landlord may in its absolute discretion from time to time determine as being that in which the accounts of the Landlord either generally or relating to the Building shall be made up

3. A copy of the Certificate prepared for each such financial year shall be supplied by the Landlord to the Tenant for approval by and without charge to the Tenant

4. The Certificate shall contain a full summary of the Landlord's said expenses and outgoings incurred by the Landlord during the financial year to which it related and the Certificate (or a copy duly certified by the Landlord's Accountant) shall be conclusive evidence for the purposes thereof of the matter which it purports to certify (save in the case of manifest error)

5. The annual amount of the Service Charge payable by the Tenant shall be a sum of money equal to the proportion hereinbefore provided by the aggregate of the said costs fees outgoings and expenses expended or incurred or payable by the Landlord in respect of the matters set out in Part 1 of this Schedule in the year to which the said Certificate relates

6. The Tenant shall pay to the Landlord such a sum (hereinafter referred to as 'the advance payment') on account of the Service Charge for the Landlord's financial year thence next ensuring as the Landlord or its Agents shall from time to time specify at its or their reasonable discretion to be fair and reasonable and such advance payment shall be paid to the Landlord in advance by equal quarterly instalments on the usual quarter days (if required by the Landlord by Bankers' Standing Order to such Bank as the Landlord shall direct) PROVIDED that subject and without prejudice to the foregoing provisions the amount of the advance payment for the Landlord's financial year current at the date of grant hereof shall be deemed to be the sum specified in paragraph 11 of the first Schedule hereto of which the Tenant shall pay on the signing hereof a due proportion calculated from day to day in respect of the period from the date hereof down to the next ensuing quarter day

7. As soon as practicable after the end of each Landlord's financial year the Landlord shall furnish to the Tenant an account of the Service Charge

payable by the Tenant for that year due credit being given therein for the advance payment made by the Tenant in respect of the said year and upon the furnishing of such account there shall be paid by the Tenant to the Landlord the Service Charge or any balance found payable or there shall be allowed by the Landlord to the Tenant any amount which may have been overpaid by the Tenant by way of advance payment as the case may require PROVIDED ALWAYS (a) that in regard to the first payment the Service Charge shall be duly apportioned in respect of the said period referred to in paragraph 6 hereof (b) that the provisions of this paragraph shall continue to apply notwithstanding the expiration or sooner determination of the term granted by the Lease so far as it relates to the demised premises but only in respect of the period down to such expiration or sooner determination as aforesaid

INDEX